Hands-On Math and Literature with MathStart®

Level 3

by Don Balka
and Richard Callan

Printed in the United States of America.

This book is printed on recycled paper.

Order Number 2-197
ISBN 978-1-58324-239-1

D E F G H 13 12 11 10 09

395 Main Street
Rowley, MA 01969
www.didax.com

CONTENTS

ABOUT THE AUTHORS

Don S. Balka, Ph.D., is a noted mathematics educator who has presented numerous workshops on the use of mathematics manipulatives with elementary, middle and high school students at national and regional conferences of the National Council of Teachers of Mathematics, at state conferences and at in-service training for school districts throughout the world. He is a former junior high and high school mathematics teacher, and is currently a Professor in the Mathematics Department at Saint Mary's College, Notre Dame, Indiana.

Richard J. Callan has been a public school teacher in Indiana for 25 years and holds his BS and MS from Indiana University. He conducts staff development workshops and makes presentations on children's literature, assessment and manipulatives. In 1995, he received the Presidential Award for Elementary School Mathematics and appears in *Who's Who in America* and *Who's Who Among America's Teachers.* He was a contributing author to the *Indiana Mathematics Proficiency Guides* in 1991 and 1997. He also was Program Chair for the NCTM Central Regional Conference, January of 2003. Rick also was the mathematics representative for the SEPA (Society of Elementary Presidential Awardees) and co-authored the books *Math, Literature and Unifix* and *Math, Literature and Manipulatives* with Dr. Don Balka.

INTRODUCTION

Integrating children's literature into your mathematics program can be a fresh, enriching experience for you as a teacher and for your children in learning mathematics, for appreciating mathematics in varied settings, and for understanding mathematics in a non-threatening, inviting environment. In *Hands-On Math and Literature with MathStart,* more fun and often challenging activities are provided to supplement those in books.

With his innovative MathStart series that includes books in three different levels, well-known children's author Stuart Murphy gives children a unique way to understand and develop the mathematics content. Each core topic selected by Murphy for his books correlates with the National Council of Teachers of Mathematics Principles and Standards for School Mathematics (2000). In many cases, the books are appropriate matches for local and state standards. The levels are by ages:

- Level 1: Ages 3 and up
- Level 2: Ages 6 and up
- Level 3: Ages 7 and up

Each level of the MathStart series examines various core topics in 21 different books. The readability of each book differs and should not be the determining factor in using the book for a specific grade level. Besides regular classroom students, special needs and ESL students will benefit from their teachers using books in the series as instructional tools or as reinforcements for concepts taught in class. Extensions of the mathematics presented in several of the books are appropriate for middle and high school students.

For each book in the MathStart series, *Hands-On Math and Literature with MathStart* presents the following pertinent information for teachers:

- Title
- Story Summary
- Grade Level
- Concepts or Skills
- Objectives
- Materials Needed
- Activities
- Writing Activities
- Internet Sites
- Assessment Ideas

Books in MathStart, Level 3, cover the following topics: Estimating, classifying, equivalent values, dividing, place value, time, percentage, angles, fractions, bar graphs, negative numbers, counting coins, metrics, building equations, reading a schedule, capacity, solving for unknowns, subtracting 2-digit numbers, dollars and cents, multiplication and mapping.

Besides the activities Murphy suggests at the end of each book, additional activities for other mathematics concepts are provided for teachers to use to expand or extend their students' mathematical learning and understanding. Teachers will be able to use these activities to develop their own lessons or thematic units of mathematics study.

Internet sites have been listed with some book entries for teacher's perusal. Some sites are inclusive with other core topics, while other sites are specific for one topic or book.

Writing or communication activities have been presented for students to think, talk, or draw about in a class or small group situation. Some of the writing prompts will provide teachers with feedback as to whether students have understood the mathematics presented. Other writing prompts provide students with opportunities to expand their thoughts and understanding of the mathematics presented in the stories.

The assessment component will let teachers measure the understanding of the mathematics using a pencil and paper task, a performance task with manipulatives, or a writing assignment. Some Internet sites will allow teachers to assess students' understanding also.

Children's literature and appropriate activities with manipulatives can be an inviting experience for children to learn and understand mathematics. By using manipulatives in the classroom, children will be able to understand mathematical information, develop mathematical concepts beyond conventional classroom settings, independently learn and understand mathematical concepts, rejuvenate creative thinking, have an appreciation for reading, and have a focal point on problem solving strategies and using connections to everyday living.

NCTM CORRELATION

	Number & Operations	Algebra	Geometry	Measurement	Data Analysis & Probability
Betcha!	√			√	
Dave's Down-to-Earth Rock Shop		√			√
Dinosaur Deals		√			√
Divide and Ride	√				
Earth Day–Hooray!	√	√		√	√
Game Time!		√		√	√
The Grizzly Gazette	√				√
Hamster Champs			√	√	
Jump, Kangaroo, Jump!	√				
Lemonade for Sale					√
Less than Zero	√			√	
The Penny Pot				√	
Polly's Pen Pal				√	
Ready, Set, Hop!	√	√			
Rodeo Time				√	
Room for Ripley				√	
Safari Park	√	√			
Shark Swimathon	√				
Slugger's Car Wash	√	√		√	
Too Many Kangaroo Things to Do!	√	√			
Treasure Map		√		√	

BETCHA!

Story Summary

Two boys try to estimate the number of jelly beans in a jar in order to win a Planet Toys contest for two tickets to the All-Star Game. Before they estimate the number of jelly beads, the boys estimate quantities and lengths for other real-life situations. The practice helps them win the competition.

New York: Harper Collins Publishers, 1997 ISBN: 0-06-026768-2

Concepts or Skills

- Estimation of quantity and length

Objectives

- Estimate length with reasonable accuracy
- Measure length with standard and non-standard units
- Estimate quantity with reasonable accuracy

Materials Needed

- Containers
- Jelly beans
- Unifix Cubes
- Paper clips
- Supermarket ads
- Calculators
- Play money
- String
- Clear plastic cups

Activity 1

Fill a container with jelly beans of one color. Have children estimate the number of jelly beans in the container. List each child's estimate. Have children count the jelly beans, and then discuss class results.

Fill a container with jelly beans of two colors. Repeat the above activity. Discuss whether or not color makes a difference in estimating the total number of jelly beans in the container.

To give children experience in making estimates, count out 10 beans and let them make an estimate. Continue this process, using 20, 30 and 40 beans each time before making a final estimate.

Activity 2

Create an estimating area in the classroom. Each week or every other week, do one of the following types of estimation activities with your children:

- For quantity, containers with: jelly beans, straws, macaroni, spaghetti, seashells, Unifix Cubes, cereal, popcorn kernels
- For length, various lengths of: jump ropes, licorice strings, pencils, string. Give children items such as Unifix Cubes or paper clips to measure the lengths.

Discuss children's estimates. For various estimation activities, discuss how they might estimate quantity or length.

Activity 3

When estimating quantity, label clear plastic cups with ones, tens and hundreds. After making their estimates, have children count the items, placing them in the ones cup, and making necessary trades to determine the exact number of items.

Activity 4

Encourage children to go with their parents to the grocery store and estimate the amount of money that will be spent. Encourage parents to make an estimate also. Have children compare the estimates.

Activity 5

Collect supermarket ads. Give each child an ad and indicate that he or she has a certain amount of money to spend at the store. By estimating, let children determine what items they could buy at the store. Have them list the items and then find the actual cost. Were they over, under, or at the actual amount of money? Have children use calculators to determine the actual amount.

Activity 6

Set up a pretend grocery store in the classroom. Have children bring in empty food containers and make a display. Give children a pretend shopping list and let them go shopping with play money you distribute. Let them make estimates on what they can purchase. Have them pay for the items, with some children serving as cashiers to make change.

Activity 7

Create estimating teams in the classroom. Let children work in cooperative groups to estimate quantities and lengths. Give each group a special name such as the following and discuss the names: Estimating Team, Calculating Team, Assessing Team, Appraisal Team, Computing Team, Approximating Team, Guessing Team and Ballpark Team.

Activity 8

Fill a clear plastic container with various types of candy (or colored plastic straws).

- Halloween: candy corn or Indian corn
- Thanksgiving: candy corn or Indian corn
- Christmas: red and green M&Ms
- Valentine's Day: candy hearts, red hots
- Easter: small chocolate eggs

Have children estimate the number of pieces in the container. Discuss class estimates before counting the objects.

Activity 9

Distribute paper centimeter tapes to each child. Have children estimate lengths of various body parts, such as circumference of head, wrist, or ankle, length of foot or forearm. Then have children measure with the tapes. Record class results and discuss the findings.

Activity 10

Have children estimate, then measure, the cycle of time for a stoplight. Are all times the same? Why are some longer?

Writing and Communicating

After each estimating experience, have children write about how they estimate the quantity or length.

Have children write on such topics as "Why Estimate?" or "Estimating in Our Daily Lives."

Assessment

The activities provide for ongoing assessment as children become more accurate in their estimates.

Notes:

DAVE'S DOWN-TO-EARTH ROCK SHOP

Story Summary

Josh is a collector. He collects baseball cards, marbles and buttons with funny sayings on them. When his Uncle Nick sends him a rock from Hawaii, Josh decides to collect rocks. He and his best friend, Amy, go to Dave's Rock Shop. They see how Dave has organized his rocks by size, shape and other categories. Together, they go rock collecting, organizing their rocks by color. When they take the collection back to Dave, he explains where rocks can be found and how they can tell us about the past. Josh and Amy collect many rocks and divide them into several different categories. Their collection is displayed for all to see.

New York: Harper Collins Publishers, 2000 ISBN: 0-06-446729-5

Concepts or Skills

- Classification
- Patterning
- Venn diagrams

Objectives

- Classify objects by different attributes
- Construct and explain a pattern
- Create a Venn diagram for various sets

Materials Needed

- Unifix Cubes
- Attribute blocks
- Beads or buttons
- Containers
- Manila paper
- String

Activity 1

Fill small containers with different colors, sizes and shapes of beads or buttons. Divide the class into small groups and give each group a container of beads. Have them sort the beads by various attributes: color, shape, size. For older children, have them classify by more than one attribute.

Activity 2

In small groups, have children make various patterns with the beads. Have children explain their patterns to the class or draw their patterns on manila paper.

Repeat this activity using Unifix Cubes.

Activity 3

Give each group some string loops to construct a Venn diagram with two or three loops. Have children classify beads or other objects by two or three attributes.

Discuss the attributes of the objects in each part of the Venn diagram. Here is an example with two attributes.

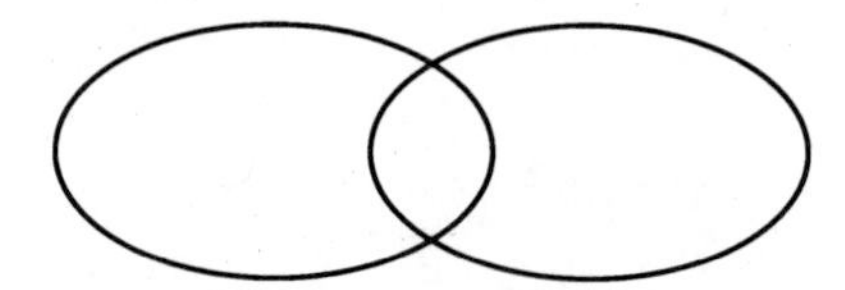

Activity 4

Repeat Activity 3 using attribute blocks.

Activity 5

Have children collect and categorize different types of rocks: sedimentary, metamorphic, igneous. Have them describe the rocks: What type? Where found?

Activity 6

Take different types of cereal (Cheerios, Chex, Frosted Miniwheats) and mix the cereals in a large container. Have children work in cooperative groups and give each group of children a scoop of cereals. Have children do the following:

- Classify the cereal.
- Make a frequency graph of their scoop of cereal.
- Share their findings with other members of the class.

Which group had the most/least of each kind of cereal?

Activity 7

Have each child take one shoe off. All children place their shoes in a pile. As a class, have children sort the shoes by one attribute, then try two attributes.

Writing and Communicating

Have children write about:

- Why do we need to classify?
- The age (or mathematics) of rocks
- When I sort...

Assessment

Give children various sets of attribute blocks and have them sort the blocks according to certain attributes.

Show sets of objects that have already been sorted according to certain attributes. Have children determine the attributes used for the sorting.

Create Venn diagrams with various objects. Have children describe the sets involved in the diagram.

Internet Link

www.rocksforkids.com

Notes:

DINOSAUR DEALS

Story Summary

Mike and his younger brother Andy have different interests, but they do have one common interest, collecting dinosaur cards. A Dinosaur Card Trading Fair was going to take place, and since it happened on Andy's birthday, his mother said he could go with Mike. Andy needs a Tyrranosaurus Rex card for his collection. This was a valuable card and required a sizable trade to acquire it. As they walked through the fair, they made other trades that were necessary to get a T-Rex. With the fair about to close, Mike finds a girl that will make the trade. However, the surprise is on Andy because the T Rex card is his birthday present from Mike.

New York: Harper Collins Publishers, 2001 ISBN: 0-06-028926-0

Concepts or Skills

- Algebra
- Equivalent values
- Graphing

Objectives

- Write equations showing equivalent values
- Complete t-tables for simple linear equations
- Construct bar graphs

Materials Needed

- Unifix Cubes
- Number cubes
- Purple, yellow, green and red 3x5 cards

Activity 1

Distribute the colored 3 x 5 cards or Unifix Cubes of the corresponding colors to students. Note that illustrations in the book make the Tyrannosaurus Rex card a dark purple and the Pterosaur card a light purple. You may want to change to a different color of card or Unifix Cube for one of the two dinosaurs.

Reread the story to students. As you read, have students show the trades on their desks. Discuss their results.

Begin symbolically writing equations to represent the trades.

3 Allosaurus = 1 T-rex	Colors: 3G = 1P
2 Triceratops = 1 Allosaurus	Colors: 2R = 1G
3 Stegosaurus = 1 Triceratops	Colors: 4Y = 1R
2 Pterosaur = 1 Stegosaurus	Colors: 2L = 1Y

Activity 2

Have students work in cooperative groups to research other dinosaurs. As they collect additional information, have each group make simple bar graphs showing the frequency of the various letters in the names of the dinosaurs. Make bar graphs of the dinosaurs described in the book also. Discuss their findings. Which dinosaur has the most letters in its name? What letter appears most? Least? Finally, make a class graph showing the results of the research.

Activity 3

To introduce coordinates to or review them with students, begin with one of the equations from Activity 1. On the overhead or chalkboard, create a two-column table marked with each color for an equation and have students determine the equivalent trades that can be made

The table below shows the equivalent values for Allosaurus (Green) and Tyrannosaurus rex (Purple) cards as given in the story. Once students have completed the table, distribute graph paper and have them plot the points. Discuss the location of the points. They lie on a straight line. If you had 0 Green Cards, how many Purple Cards could you trade for? The answer is 0. Does the point (0,0) lie on the same straight line?

Green	Purple	Point
3	1	(3, 1)
6	2	(6, 2)
9	3	(9, 3)
12	4	(12, 4)
15	5	(15, 5)
18	6	(18, 6)

Repeat the activity using different equations from Activity 1.

Activity 4

For upper grades, have students use Unifix Cubes to show equivalent relationships where two or more equations are involved. For example, if:

1G = 2R, then 3G = 6R = 1P

4Y = 1R, then 2R = 8Y = 1G and 3G = 6R = 24Y = 1P

2L = 1Y, then 8Y = 16L = 1G and 3G = 48L

Activity 5

Play Making Trades. Divide students into groups of three or four. Give each group a number cube and several Unifix Cubes of two colors, along with one Unifix Cube of a third color. Before starting the game, indicate to students how trades will be made. For example, 3 Red cubes could be traded for 1 Blue cube, and 4 Blue cubes could be traded for 1 Green cube. One student begins play by rolling the number cube. The number indicates how many cubes the student can select of the first color. Using the equivalencies above, a 6 would be used for 6 Red cubes, which could be traded for 2 Blue cubes. A 5 would be used for 5 Red cubes, which could be traded for 1 Blue cube with 2 Red cubes still remaining. To win the game, a student must get the exact number of Blue cubes to make the trade for 1 Green cube.

Change the equivalencies and have students play the game again.

Writing and Communicating

Have students write about or discuss making trades with baseball cards or other collectibles.

Have students create their own trading scheme for three or four objects. For example, 1A = 2B, 1B = 3C, 1C = 5D.

Assessment

GIven a table of equivalencies, have students plot the corresponding points on a coordinate grid.

Have students find an equivalence relationship for a given set of equations.

Internet Link

www.zoomdinosaurs.com

Notes:

DIVIDE AND RIDE

Story Summary

A group of eleven friends attend the carnival. To go on most of the rides, they must be divided into groups of two, three, or four; however, all seats must be filled. In order to do this, the friends ask unknown children to fill the empty seats. Finally, on the last ride, the eleven friends can ride together, but they still need to fill empty seats. They invite their new friends to join them for a great ride.

New York: Harper Collins Publishers, 1997 ISBN: 0-590-21427-6

Concepts or Skills

- Division
- Quotient
- Remainder
- Prime numbers

Objectives

- Model division problems with and without remainders
- Find the quotient and remainder of a division problem
- Determine prime numbers and composite numbers up to 20

Materials Needed

- Unifix Cubes
- Two-color counters
- Paper bags

Activity 1

Divide students into small groups of two, three, or four. Give each group a small bag of Unifix Cubes. Have students divide the cubes equally among members of the group, and then write a division number sentence, listing the quotient and any remainder. Have students share their number sentence with the class.

Activity 2

Give each student a handful of two-color counters. Have students write a short paragraph where division is used with the total number of counters. Have them write the corresponding division number sentence.

Activity 3

Divide students into groups of two. Give each group 20 Unifix Cubes or two-color counters. Have students number from 1 to 20. Have students make different rectangular arrays of cubes for each counting number. The arrays can be made horizontally and vertically. For each array, have them record the corresponding multiplication sentence. The results are shown in the table.

Number	Possible Arrays
1	1x1
2	1x2, 2x1
3	1x3, 3x1
4	1x4, 2x2, 4x1
5	1x5, 5x1
6	1x6, 2x3, 3x2, 6x1
7	1x7, 7x1
8	1x8, 2x4, 4x2, 8x1
9	1x9, 3x3, 9x1
10	1x10, 2x5, 5x2, 10x1
11	1x11, 11x1
12	1x12, 2x6, 3x4, 4x3, 6x2, 12x1
13	1x13, 13x1
14	1x14, 2x7, 7x2, 14x1
15	1x15, 3x5, 5x3, 15x1
16	1x16, 2x8, 4x4, 8x2, 16x1
17	1x17, 17x1
18	1x18, 2x9, 3x6, 6x3, 9x2, 18x1
19	1x19, 19x1
20	1x20, 2x10, 4x5, 5x4, 10x2, 20x1

Array Examples:

x x x	x x x x x
x x x	x x x x x
2 x 3	x x x x x
	3 x 5

Numbers that can be represented by exactly two arrays are called prime numbers: 2, 3, 5, 7, 11, 13, 17, 19.

Numbers that are represented by more than two arrays are called composite numbers: 4, 6, 8, 9, 10, 12, 16, 18, 20.

The number 1 is neither prime nor composite.

Activity 4

Play Target Divide. Give each student a target number and the corresponding number of Unifix Cubes or two-color counters. Have each student divide the target number by 2, 3, 4 and 5, writing the division number sentence showing the quotient and remainder. If the target number is 13, for example, then:

$13 \div 2 = 6$, R 1	$13 \div 4 = 3$, R 1
$13 \div 3 = 4$, R 1	$13 \div 5 = 2$, R 3

Discuss the number sentences and the possible remainders for each divisor. For example, when a number is divided by 5, the possible remainders are 0, 1, 2, 3, or 4

Writing and Communicating

Have each student create his/her own story involving friends and rides at a fair, carnival, or Disneyland.

Have students respond to "The remainder is [N] when any counting number is divided by 1, 2, 3, 4, 5, 6, 7, 8, or 9." For example, if a counting number is divided by 7, the possible remainders are 0 through 6.

Assessment

Have students model with Unifix Cubes or two-color counters a given division problem.

Have students determine if a given number is prime or composite using cubes or two-color counters.

Internet Links

kids.msfc.nasa.gov

www.edu4kids.com

www.aplusmath.com

Notes:

EARTH DAY–HOORAY!

Story Summary
The Maple Street School Save-the-Planet Club decided to clean up Gilroy Park for the Earth Day celebration. By collecting and recycling 5000 cans, they could get enough money to buy flowers to plant at the park. As students pick up cans, they place them in bags of 10. When they had 10 bags of cans, they placed them in one bag that represented 100 cans. The students continued with their project. When other students and the community become involved, the Club reaches and exceeds its goal. Now 10 bags of 100 are placed in an even larger bag.

New York: Harper Collins Publishers, 2004 ISBN: 0-439-74908-5

Concepts or Skills
- Place value
- Expanded notation
- Measurement
- Proportions
- Problem solving
- Graphing
- Algebra

Objectives
- Understand place value through 1000
- Write a number in expanded notation
- Determine the weight of objects
- Solve proportions
- Graph data
- Estimate quantities

Materials Needed
- Unifix Cubes
- Number tiles 0-9
- Two-color counters
- Plastic containers
- Macaroni, rocks, candy, or M&Ms
- Shark Base 10 Pieces, page 61
- Six-sided number cubes

Activity 1
Start a recycling club at your school. Place boxes or plastic barrels in each classroom and have children and teachers deposit any materials that can be recycled. The club may want to decorate the boxes or barrels to encourage children and teachers to recycle paper.

Contact you local recycling center to determine what types of items they will recycle and how they should be packaged. Have a contest to determine which grade level or classroom can bring in the most recyclable items.

Find the current recycling rates for aluminum and for paper. Have children regularly determine the weight of recycled materials using proportions and determine the amount of money saved.

Activity 2
Place a set of number tiles 0 through 9 in a paper bag and distribute a set to a pair of children. Have one child draw 3 or 4 tiles from the bag, depending on grade level. The other child (or cooperatively):
- Makes the greatest number with the tiles. Discuss the place value of each digit.
- Makes the least number with the tiles. Discuss the place value of each digit.
- Makes an equation with the digits using addition or subtraction.

Partners continue this activity replacing the tiles for each new draw. Discuss the greatest possible number that could be created (987; 9876).

Discuss the least possible number that could be created (102; 1023). Extend the problem to 5 or more digits and have children note any patterns.

Activity 3

Fill a larger plastic container with any of the following items: candy, small rocks, M&Ms, buttons, macaroni. Let children estimate the number of objects in the container, writing their estimates on a note card. Use different sizes of plastic bags to group the objects, small bags for 10, larger bags for 10 small bags. Have children write the final result in expanded form.

Activity 4

Give each pair of children two number cubes. Each child tosses the cubes and attempts to make the greatest number possible or the least number possible. Children can decide which way to play the game. After each round of tossing the cubes, the child with the greater number earns one point and records the point. After 5 (10) rounds, the winner is the child with the most points. Using the same rules as above, have students take 3 or 4 dice and repeat the activity. Have children record their numbers on paper for each round.

Activity 5

To begin, toss a pair of number cubes, record the two digits on the overhead or chalkboard, and then discuss how many two digit numbers can be created with the two digits. There are only two. For example, if a 3 and 5 are tossed, then the only two numbers that can be constructed are 35 and 53.

Give each pair of children three number cubes. Have one of the children toss the three cubes. With the three numbers showing, each child writes all the possible 3-digit numbers. Discuss class findings. When three cubes are tossed, there are 6 possible 3-digit numbers that can be constructed. For example, if 1, 2 and 3 are tossed, then the six numbers are 123, 132, 213, 231, 312 and 321.

Continue the activity by adding a fourth number cube. Have children look for patterns. With four digits, there are 24 different 4-digit numbers that can be constructed. For example, if 1, 2, 3 and 4 are tossed, here are the numbers that can be constructed:

1234	2134	3124	4123
1243	2143	3142	4132
1324	2314	3214	4213
1342	2341	3241	4231
1423	2413	3412	4312
1432	2431	3421	4321

What is the pattern? Look at the table below.

Number of Digits	Number of Numbers
1	1 = 1 = 1!
2	2 = 1 x 2 = 2!
3	6 = 1 x 2 x 3 = 3!
4	24 = 1 x 2 x 3 x 4 = 4!
5	120 = 1 x 2 x 3 x 4 x 5 = 5!
.	.
.	.
n	1 x 2 x 3 x 4 x ... x n = n!

The notation shown in the table is called factorial notation. It is notation used in algebra, probability and statistics.

Activity 6

Use the Base 10 pieces from Shark Swimathon. Each child should have at least ten unit squares, ten longs and five 100-squares. Call out a number, such as 53, and have children:

- Write the number in Standard Form (53).
- Write the number in Expanded Form (50 + 3 or 5 x 10 = 3 x 1).
- Show the number using the paper place value models.

The teacher can assess children's understanding of the ideas.

Writing and Communicating

Have children write or tell about what happens when two digits are switched in a two-digit number.

Have children write about how our place value system helps us to understand numbers.

Have children research and write about place value in ancient numeration systems, such as the Mayans, Babylonians, Romans and Greeks.

Assessment

Give a child a set of number tiles 0 through 9. Orally state a 3-digit or 4-digit number. Have the child construct the number with the tiles.

Construct a 4-digit number with the tiles. Have the child pull down the tile that is in a particular position.

GAME TIME!

Story Summary

In one week, the Huskies soccer team will be playing the defending champion Falcons soccer team for the championship. As they practice during the week, the girls use different time units to describe current or future events. The day of the game arrives. With only a few seconds remaining, the Huskies score a goal to win the championship.

New York: Harper Collins Publishers, 2000 ISBN: 0-06-028024-7

Concepts or Skills

- Measuring time
- Variables
- Bar graphs

Objectives

- Understand time units
- Calculate equivalent time units
- Understand the concept of variable
- Determine number patterns
- Construct a bar graph

Materials Needed

- Weekly and monthly calendars
- Overhead clock faces

Activity 1

Distribute overhead clock faces to each child. Orally call out a time and have children construct the time on their clocks.

Activity 2

Form groups of two children and distribute a monthly calendar to each group. Have children look for number patterns on the calendar. Discuss group findings. Do the same patterns occur for different months?

Here are some possible patterns.

- In going down a column on a calendar, there is a difference of 7 between two consecutive dates.
- In a square cluster of four days, the sums of the numbers on diagonals are equal. An illustration is shown below:

  6 7

  13 14 Note: $6 + 14 = 7 + 13$
- The difference between the lower right and upper left numbers is always 8, whereas the difference between the lower left and upper right numbers is always 6.
- In a square cluster of nine days, the sum of the numbers in the cluster is nine times the middle number. An illustration is shown below:

  1 2 3

  8 9 10

  15 16 17 Note: Sum $= 81 = 9 \times 9$
- In a square cluster of nine days, the sums of the numbers on each diagonal are equal.

Activity 3

In finding the patterns described in Activity 2, simple ideas of algebra can be introduced to children. Here are the equivalent algebraic ideas described.

- In a square cluster of four days, let the first day be represented by N. Then the cluster has the following form:

N	N + 1
N + 7	N + 8

  On the diagonals,
  N + N + 8 = N + 1 + N + 7 = 2N + 8.

  The differences on the diagonals will be
  N + 8 - N = 8 or N + 7 - (N + 1) = 6.

  The sum of the numbers in the cluster is
  N + N + 1 + N + 7 + N + 8 = 4N + 16
- In a square cluster of nine days, let the middle number be represented by N. Then the cluster has the following form:

N - 8	N - 7	N - 6
N - 1	N	N + 1
N + 6	N + 7	N + 8

  On the diagonals,
  N - 8 + N + N + 8 = 3N = N + 6 + N + N - 6.

  The sum of the nine numbers is 9N, which is 9 times the middle number.

Activity 4

There are many equivalent times presented in the story. Discuss the various equivalencies. Have children determine the number of equivalent units for various time periods. Some possibilities are listed below.

- 1 week = 7 days = 7 x 24 hours = 7 x 24 x 60 minutes = 7 x 24 x 60 x 60 seconds
- 1 year = 365 days (non-leap year) = 365 x 24 hours = 365 x 24 x 60 minutes

Activity 5

Give children calculators. Have them write their age to the nearest year and enter the age on the calculator. Have them determine their ages in months, days, hours, minutes and seconds. There are several ways to calculate the answers. Variations in answers will occur depending on which units are used.

For example:

10 years = 10 x 12 months

10 years = 10 x 365 days

10 years = 10 x 365 x 24 hours

Activity 6

Using class data for birth months, have children make bar graphs.

Writing and Communicating

Have children write about time around the world.

Have children write about where time "starts" at Greenwich, England.

Have children research and write about why standard units of time were necessary.

Assessment

Give children a different type of number chart, such as the one illustrated below. Have them determine various patterns.

1	2	3	4	5
6	7	8	9	10
11	12	13	14	15
16	17	18	19	20
21	22	23	24	25

Similar patterns occur on this chart as those described in the activities.

Have children determine a given time unit in an equivalent unit. For example, 10 minutes = 600 seconds.

Internet Links

www.calendarhome.com

www.time.gov

Notes:

THE GRIZZLY GAZETTE

Story Summary

With only one week left at Camp Grizzly, a camp mascot must be elected to wear the Grizzly bear outfit and lead the parade. Two candidates have already begun their campaigns, but Jacob convinces Corey that she could still win. Daily polls of the 100 campers, with results presented as percents in circle graphs, show the voting patterns.

On the last day, each candidate gave a final speech. Sophie did cartwheels, Daniel reminded them of the candy bars he gave out, but Corey wrote a new camp cheer which the Music Club played. The final results show Corey winning the election with 50 percent of the votes.

New York: Harper Collins Publishers, 2003 ISBN: 0-06-000027-9

Concepts or Skills

- Percents
- Circle graphs

Objectives

- Construct a circle graph for given percents
- Determine equivalent percents for given fractions
- Collect data and construct an appropriate circle graph

Materials Needed

- 2 cm Grid Paper, page 48
- Blank Circle Graph, page 49
- Fraction Cards (4 copies), page 50

Activity 1

Discuss the definition of percent (per one hundred). Look at each of the circle graphs illustrated in the book. Since there were exactly 100 campers, the percents throughout the story actually give the number voting for each candidate. Change the number of campers to 200, and introduce the change by saying "What if there were 200 campers. If 50% of 200 had not made up their minds, how many campers does that represent?" Continue with the other percentages presented. Discuss whether the circle graphs change. Note that the sum of all the percents is 100%. Note that 25% is 1/4 of the circle, 50% is 1/2 of the circle, and 75% is 3/4 of the circle.

Activity 2

Have children work in groups of two, each with a 2 cm Grid Sheet. Give each group a bag containing 10 Unifix Cubes with three or four different colors of cubes (for example, 2 Red, 3 White, 4 Green, 1 Orange). Using the idea of percent, have children color the grid sheet to correctly denote the percents. As a class, discuss the findings for each group. Was the grid correctly colored?

Activity 3

As suggested at the end of the story, have students conduct polls with classmates, family, or friends. Limit the sampling to 10 persons. Present different options to students for the poll: favorite sport, favorite color, favorite candy, favorite food, favorite TV show. For each of these, however, provide a list of choices.

Once students have conducted their surveys, have them determine the corresponding percents and then construct an appropriate circle graph. Compare graphs of students that selected the same option. Note the differences in the survey results.

Post the circle graphs on a bulletin board.

Activity 4

Distribute a set of Fraction Cards to groups of 3 or 4 students. Each student needs a Circle Graph and crayons. The set of cards are placed face down. One student starts by drawing a card and coloring the corresponding area on the Circle Graph. Play continues in the same manner. When one student draws a card that results in over 100% of the graph being colored, the game is over. The student who is closest to 100%, without going over, is the winner.

Writing and Communicating

Have children write their own stories about camp that involve percents.

With a given set of percents, present a different "What if there were" and have students write about how they determined the numbers represented by each percent.

Assessment

Have students determine percents for a given set of data.

Have students construct a circle graph for a given set of percents.

Have students color a 10 x 10 grid for a given set of percents.

Notes:

HAMSTER CHAMPS

Story Summary

Chuckles, Moe and Pipsqueak are hamsters left alone with Hector the cat. They propose to show him new stunts if he promises not to chase them. They find a protractor to measure angles and form a ramp with a 30 degree incline. They jump into a toy car going down pillows from the couch and hit the ramp going fast. The car soared into the air with the hamsters aboard. Hector was not impressed, so the hamsters continued their stunts by changing the angles.

New York: Harper Collins Publishers, 2005 ISBN: 0-06-055772-9

Concepts or Skills

- Measuring angles

Objectives

- Measure an angle with a protractor
- Identify acute, obtuse and right angles
- Draw an example of an acute, obtuse, or right angle
- Describe a straight angle as an angle that has a measure of 180°

Materials Needed

- Protractors
- Geoboards and rubber bands
- Tangrams
- Pattern blocks
- 50 centimeter lengths of string or yarn

Activity 1

Give each child a clock face with hands and a protractor. Have children set the following times on their clocks and then measure the angles formed using the protractor

- 1 o'clock 30 degrees
- 2 o'clock 60 degrees
- 3 o'clock 90 degrees
- 4 o'clock 120 degrees
- 5 o'clock 150 degrees
- 6 o'clock 180 degrees
- 7 o'clock 210 degrees
- 8 o'clock 240 degrees
- 9 o'clock 270 degrees
- 10 o'clock 300 degrees
- 11 o'clock 330 degrees
- 12 o'clock 360 degrees

As students measure, check for consistency in their answers. What do they notice about their findings? The hour hand moves 30 degrees in one hour.

Activity 2

Group children in threes and give each group a 60 centimeter length of string or yarn. Have one child find and hold with one hand the middle of the string. This will be the vertex of an angle. Have the other two children hold the ends of the string. Direct children to form a 90° angle, a 45° angle, an obtuse angle, an acute angle, a straight angle. Challenge children to form a 30° or 60° angle. Have another group measure the angle with a protractor.

Activity 3

Give each child a geoboard and some rubber bands. Have children make various types of angles on the geoboard. Use an overhead geoboard to model their results. Have children use a protractor to measure the angles.

Activity 4

Give each child some pattern blocks. Have them determine the interior angle measures of all the blocks.

Orange Square:	90, 90, 90, 90
Green Equilateral Triangle:	60, 60, 60
Blue Rhombus:	60, 120, 60, 120
Red Trapezoid:	60, 120, 120, 60
Yellow Hexagon:	120, 120, 120, 120, 120, 120
Tan Rhombus:	30, 150, 30, 150

Activity 5

Give each child a set of tangrams. Have them determine the interior angle measures of the seven pieces.

- There are five isosceles right triangles of different sizes: 45, 45, 90.
- There is one square: 90, 90, 90, 90.
- There is one parallelogram: 45, 135, 45,135

Activity 6

Have children stand and face one direction(north). Then ask them to face east. What degree of an angle did they turn? (90) Then, have them turn south. From their original direction (north) what angle have they now turned through? (180) Continue this process with the following:

a. Turn clockwise northeast (NE):	45
b. Turn clockwise northwest (NW):	315
c. Turn clockwise southeast (SE):	135
d. Turn clockwise southwest (SW):	225

Writing and Communicating

Have children write about right angles, acute angles, and obtuse angles that are in the classroom.

Have children write about buildings that have acute angles or obtuse angles.

Mathematical Notes

An angle is defined as the union of two rays with a common endpoint. The endpoint is called the vertex.

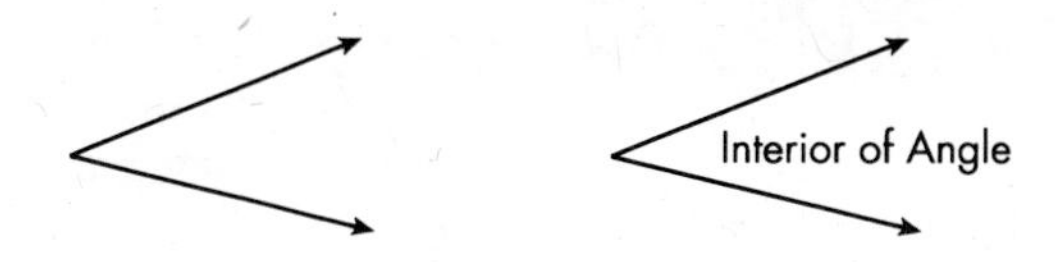

- An angle is measured in degrees. 1 degree = 1/360 of a circle.
- An acute angle has a measure greater than 0 degrees and less than 90 degrees.
- A right angle has a measure of 90 degrees.
- An obtuse angle has a measure greater than 90 degrees and less than 180 degrees.
- Some textbooks define a straight angle as an angle that has a measure of 180 degrees.

Assessment

Give each child a card with acute and obtuse angles. Have them identify the angles, estimate the measure, and then measure with a protractor.

Notes:

JUMP, KANGAROO, JUMP!

Story Summary

Twelve campers, including a kookaburra, an emu, platypuses, koalas, dingoes and a kangaroo, participate in a field day run by the kangaroo camp counselor Ruby. The first activity pits one half of the campers against the other half in a tug-of-war. Kangaroo's team loses. For the swimming relay race, campers split into thirds. Kangaroo's team came in second. In the third contest, a canoe race, campers were divided into fourths. Kangaroo's team tied for last place. Kangaroo is distraught for not being on a winning team. The last contest is an individual one, the long jump. Kangaroo wins and sets a new camp record.

New York: Harper Collins Publishers, 1999 ISBN: 0-06-027614-2

Concepts or Skills

- Fractions
- Equivalent fractions
- Addition of fractions

Objectives

- Determine and write equivalent fractions
- Determine fractional parts of time units
- Add like fractions

Materials Needed

- Unifix Cubes
- Pattern blocks
- Two-color counters
- Plastic coins
- Large display clock

Activity 1

Distribute 12 Unifix Cubes or two-color counters to each student.

Reread the story to students. As you read, have students show with the cubes on their desks how the campers are divided into teams. Discuss the idea of equivalent fractions. Each camper represents 1/12. For the first contest, 1/2 = 6/12; for the second contest, 1/3 = 4/12; for the third contest, 1/4 = 3/12.

Activity 2

Distribute 16 Unifix Cubes or two-color counters to each student.

Reread the story to students using 16 campers, making four teams for swimming and 8 teams for canoeing. As you read, have students show with the cubes on their desks how the campers are divided into teams. Discuss the idea of equivalent fractions. Now, each camper represents 1/16 of the total. For the first contest, 1/2 = 8/16; for the second contest, 1/4 = 4/16; for the third contest, 1/8 = 2/16.

Activity 3

Using a large display clock, show students how fractions play an important role in talking about time measures. Show how a dial can be divided into quarters or halves. For older students, other fractional parts of the clock face could also be shown. Pose questions such as the following:

- How many minutes in 1 hour? 60
- One minute would be what fraction of 1 hour? 1/60

- How many minutes in 1/4 of an hour? 15 minutes
- What fraction is equivalent to 1/4? 15/60

Many more are possible: 1/2 hour, 3/4 hour, 1/3 hour, 2/3 hour, 1/6 hour.

With both hands on 12, have individual students show various times on the clock.

Activity 4

Using plastic or real coins, have students display various fractional parts of one dollar or fractional parts of other coins.

- 1 cent = 1/5 of a nickel
  1 cent = 1/10 of a dime
  1 cent = 1/25 of a quarter
  1 cent = 1/50 of a half dollar
  1 cent = 1/100 of a dollar
- 1 nickel = 1/2 of a dime
  1 nickel = 1/5 of a quarter
  1 nickel = 1/10 of a half dollar
  1 nickel = 1/20 of a dollar
- 1 dime = 1/5 of a half dollar
  1 dime = 1/10 of a dollar
- 1 quarter = 1/2 of a half dollar
  1 quarter = 1/4 of a dollar

Discuss other fractional parts.

Activity 5

Divide students into small groups and give each group a container of pattern blocks. Have them make hexagons using each color: green, blue and red. First, make hexagons with all the same color of pattern blocks. Discuss equivalencies.

1 yellow hexagon = 2 red trapezoids = 3 blue rhombi = 6 green equilateral triangles

A green triangle is 1/6 of a yellow hexagon. A blue rhombus is 1/3 of a yellow hexagon. A red trapezoid is 1/2 of a yellow hexagon.

After students have experienced making same-color hexagons, have them make hexagons using combinations of pieces. Introduce addition sentences, such as the following: 1/2 + 1/3 + 1/6 = 1 (red trapezoid+ blue rhombus + green triangle = yellow hexagon).

Writing and Communicating

Have students write about or discuss why we need fractions in our daily lives.

Have students create their own "field day" story with various contests and a specified number of campers.

Assessment

Give students a specified number of Unifix Cubes or two-color counters. Have them divide the cubes into a given fractional part (1/2, 1/3, 1/4).

Give students fraction circles divided into twelfths. Have students shade various fractional parts.

Internet Links

www.kidshealth.org/kid

www.coolmath.com

Notes:

LEMONADE FOR SALE

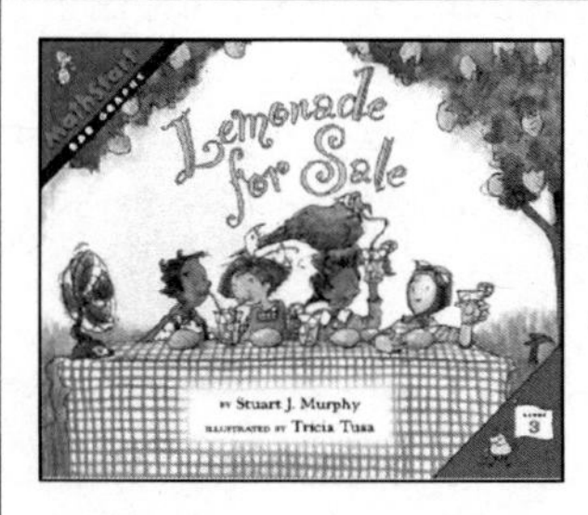

Story Summary

The Elm Street Kids Club needs to make repairs on its clubhouse. The kids decided to sell lemonade to make money in order to make the repairs. They keep track of their sales by using a bar graph to show how much they sell each week.

New York: Harper Collins Publishers, 1998 ISBN: 0-06-446715-5

Concepts or Skills

- Data analysis
- Graphing
- Range
- Mean

Objectives

- Construct a bar graph
- Interpret a bar graph
- Determine the range for a set of data
- Determine the mean for a set of data

Materials Needed

- Unifix Cubes
- Pennies
- Bulletin board paper
- Name Graph Page, page 51
- 2 cm Grid Paper, page 48

Activity 1

Have each student bring two pennies to class, or provide them for each student. Label bulletin board paper with the title "Pennies Represented by Decade." List the following dates on the paper: 1960, 1970, 1980, 1990, 2000. Have students check the date on each penny and then write the year on the graph for the corresponding decade. For example, if the penny date is 1978, the student will write "1978" on the line corresponding to 1970.

Discuss the resulting graph. Which decade had the most(least)pennies? Are there decades that had the same number of pennies?

Activity 2

Label bulletin board paper with the title "Favorite Subject in School." List the following subjects on the paper: spelling, mathematics, social studies, science, reading, other. Have students mark their favorite subject with an "x." Discuss the resulting graph.

Activity 3

Make copies of the Name Graph page. Cut apart and distribute a strip of squares to each student. Have students print their first (last) names on the strip, one letter in each square. Have them cut off any blank squares. On a bulletin board, mark a horizontal axis with numbers 2, 3, ..., 9, 10, concluding with the word "More." Have each student tape their name strip above the corresponding number representing the number of letters in their name. Discuss the resulting graph.

Activity 4

Make an Our Favorite Foods graph! Give each student a piece of 2 grid paper. Assign a particular food or drink type to each student or pair of students. Depending on grade level, axes may already be labeled with information. Have students collect class data and make a bar graph. Display the graphs and then discuss the results.

Here are possible food types: meats, vegetables, fruits, cookies, desserts, candy bars, drinks. Also include "Foods we dislike."

Writing and Communicating

Have students respond to the prompt "How We Can Learn from Making and Using a Bar Graph."

Assessment

Present students with survey data. Have them construct a bar graph showing the data. Have students write about the data.

Notes:

LESS THAN ZERO

Story Summary

Perry the penguin is trying to save for an ice scooter that will cost him 9 clams. Perry has no clams and so he does chores around the house, but spends the clams he earns on other things. He soon realizes that he doesn't have enough clams and he owes clams to his friends. For one week, he keeps track of his clams using a line graph to show how he is doing on his saving. On some days, Perry has less than zero, but eventually, he earns enough to purchase his scooter.

New York: Harper Collins Publishers, 2000 ISBN: 0-06-028034-4

Concepts or Skills

- Line graphs
- Negative integers
- Number lines

Objectives

- Construct a number line
- Use a number line involving negative integers

Materials Needed

- Paper
- Pencil
- Clam Number Line, page 52
- Unifix Cubes
- Less than Zero Story Graph, page 53
- Less than Zero Week Graph, page 54

Activity 1

Distribute a Clam Number Line and one Unifix Cube to each child. As suggested in the book, have children keep track of Perry's clams using a marker on the number line. Have children place their cube on 0. As Perry gains clams, have them count to the right and place the cube on the number line. As he spends or loses clams, have them count to the left and place the cube on the number line. Have children write the corresponding number sentences in the story: $0 + 4 = 4$, $4 - 5 = -1$, $-1 - 2 = -3$, $-3 + 8 = 5$, $5 - 8 = -3$, $-3 + 8 = 5$, $5 + 4 = 9$.

Create other clam stories. Have children mark locations on the Clam Number Line and write the corresponding number sentences.

Activity 2

Make a transparency of the Less Than Zero Story Graph. Reread the story as suggested, counting the clams and showing the location on the graph.

Use the clam stories from Activity 1 to plot points and graph on the Less Than Zero Week Graph.

Activity 3

Using the Less Than Zero Week Graph, have children select a city that will have subzero temperatures during the winter months. Have them record the daily temperatures for one week. Use a variety of cities. Post the graphs and discuss the findings. Students will need to determine a scale for temperatures on the vertical axis.

Writing and Communicating

Have children explain how a graph can help them understand their findings.

Have children write their own "week" story about earning and spending money.

Assessment

Present children with various "week" graphs. Ask questions dealing with the graph. Most? Least? Less than zero? Greater than zero? Zero?

Notes:

THE PENNY POT

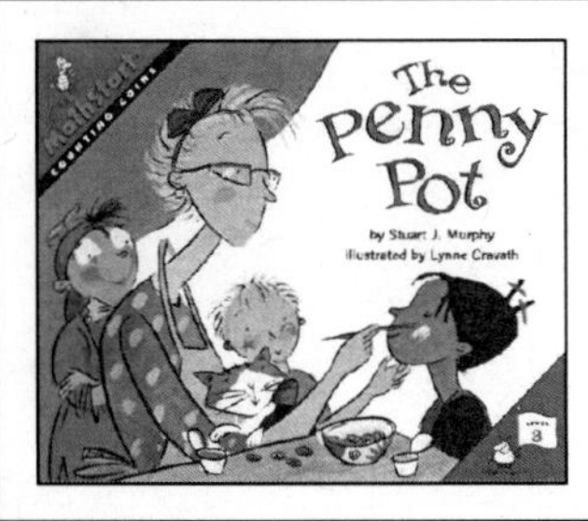

Story Summary

Jessie is at the school fair and would like to have her face painted. However, she does not have enough money. At the face painting booth, there is a penny pot where people place their extra pennies. As Jessie's friends get their faces painted, they place pennies in the pot. Eventually, there is enough money for Jessie to get her face painted also.

New York: Harper Collins Publishers, 1998 ISBN: 0-06-027607-X

Concepts or Skills

- Coin values
- Problem solving

Objectives

- Identify coins
- Determine the value of a given collection of coins
- Construct a table to determine solutions for coin problems

Materials Needed

- Pennies
- Other coins or plastic coins
- Calculators
- Coin Mat, page 55

Activity 1

Give each child one penny. Have them observe both faces and discuss what they see. Make a people graph or object graph for the dates on the pennies. Discuss results of the graph.

Activity 2

Fill a container with pennies. Have children estimate the number of pennies in the container. Count the pennies and discuss how to write the amount in dollars and cents.

Activity 3

Place a certain amount of change in your pocket or hand. For example, tell children that you have 1 quarter, 3 dimes and 4 nickels. Have children use plastic coins or paper cut-outs to determine how much money you have. Continue the activity throughout the school year.

Activity 4

Discuss the different coins: penny, nickel, dime, quarter, half dollar. Tell students that you have a certain number of coins in your hand and ask them how much money you could have. As they respond, write their sums on the board or overhead projector.

Discuss how they might organize all of their responses in a table. What is the smallest or largest amount of money that you could have?

Repeat the activity with a different number of coins.

Activity 5

In the back of the book, Stuart Murphy has devised a game called "Trading Coins." Use the game board and have students play the game.

Activity 6

Have students bring in empty food containers and make a store. Let children go to the store with various amounts of money to buy different items, with other children acting as cashiers.

Activity 7

Take old catalogs and let children do Christmas or birthday shopping for their family using the catalogs. Give each child a set amount of money to spend. Children can use calculators to determine how much is spent. After the activity, discuss who spent the most and least.

Activity 8

The U.S. Mint is now producing state quarters in the order in which the states were admitted to the Union. Show the new quarters and discuss their obverse faces.

Activity 9

Distribute a Coin Mat to each child. Distribute a number cube and several real or plastic coins or picture cut-outs of coins to each group of two or three children. Include lots of pennies and other coins through a quarter.

On a turn, a child rolls the number cube, selects the corresponding number of pennies and places them in the column on the Coin Mat. If possible, the child makes trades (1 nickel for 5 pennies, 1 dime for 2 nickels, 1 quarter for 2 dimes and 1 nickel.)

The first player to get exactly 1 quarter is the winner. If a child cannot make 1 quarter exactly, he or she loses that turn.

Writing and Communicating

Have children respond to prompts, such as the following:

- I have five coins: pennies, nickels and dimes. How much money could I have?
- I have five coins: dimes and quarters. How much money could I have?

Have children write about the following prompts:

- The Traveling Coin
- Going from Pocket to Pocket

Internet Links

www.younginvestor.com

www.kidsbank.com

www.kokogiak.com/megapenny

Assessment

The writing assignments mentioned above serve as good reinforcement tools. Students not only need to find the value of the given number of coins, but also to use problem solving strategies to determine all possibilities.

Notes:

POLLY'S PEN PAL

Story Summary
Polly and Ally are pen pals. As they communicate through email messages, Polly learns about the metric system that is used in Canada. Her father helps her by suggesting comparable items or distances that she can understand. When her father goes on a business trip to Montreal, Canada, Polly goes with him and gets to meet Ally.

New York: Harper Collins Publishers, 2005 ISBN: 0-06-053168-1

Concepts or Skills

- Comparing English and Metric measures

Objectives

- Estimate given lengths in centimeters
- Estimate given weights in grams
- Read a metric ruler in centimeters
- Compare measurements in English units and metric units

Materials Needed

- Unifix Cubes
- String
- Liter bottle
- Measuring cup
- Water
- Metric scale
- Metric balance scale and weights

Activity 1

Give each child a string that is 10 centimeters in length. Have children take the string home and find 5 items that measure close to 10 centimeters. Have them write down the items and then return to school with their list. Discuss the items in class.

Activity 2

Have children write the to the web site listed on the next page to get a pen pal that lives in a country that uses the metric system. Have children ask questions about their pen pals that relate to the metric system. Let children share their letters with the class as they receive them.

Activity 3

Take an empty liter container and a measuring cup. Let children determine the approximate number of cups it would take to make a liter. First, have children estimate the number of cups.

Activity 4

Conduct a metric scavenger hunt as suggested in the book. Give each child a metric ruler. Have them find five objects in the classroom that are approximately 20 centimeters in length. Allow for lengths between 18 and 22 cm.

The first person to find the five objects is the winner. Discuss in class the various items that were measured. Determine if children measured the same items and if they recorded the same lengths.

Activity 5

Using a metric scale, have children find their weights in kilograms. Using an English scale, have children find their weights in pounds. List the two weights as an ordered pair (metric, English). Have children make a scatterplot of the class data. Discuss the graph.

Using a meter stick, have children find their heights in centimeters. Using a yardstick, have children find their heights in inches. List the two heights as an ordered pair (metric, English). Have children make a scatterplot of the class data. Discuss the graph.

Activity 6

A nickel weighs about 5 grams. Have children hold a nickel. Twenty nickels would weigh 100 grams, and 200 nickels would weigh 1000 grams or 1 kilogram. Have children hold the 200 nickels. Large grapefruit weigh about 1 kilogram. Have children hold the coins in one hand and a grapefruit in the other to compare weights.

Activity 7

Give each child a centimeter ruler. Have children measure to the nearest centimeter the length and width of a book, the length and width of their desk, or the length of a particular item they all might have at their desks.

Writing and Communicating

Have children write about how the metric system could be implemented in our country.

Have children write about the advantages (disadvantages) of the metric system.

Assessment

On sheets of paper, draw line segments of various lengths. Have children estimate the lengths in centimeters. Then have them measure with a ruler.

Have children weigh various objects using a metric balance scale.

Internet Links

www.kids-space.org

www.k111.k12.il.us/king/math.htm

www.epals.com

Notes:

READY, SET, HOP!

Story Summary

Matty and Moe are two frogs that like to hop. They engage in a hopping contest, first to a rock, then to a log, and finally to a pond. Each wins a segment of the contest, and both eventually end up in the pond with the same number of hops.

New York: Harper Collins Publishers, 1996 ISBN: 0-06-025877-2

Concepts or Skills

- Addition
- Subtraction
- Pre-algebra

Objectives

- Calculate sums with three addends
- Write basic addition number sentences
- Construct a frequency bar graph
- Calculate appropriate statistics for a set of data

Materials Needed

- Unifix Cubes
- Two-color counters
- Plastic cup
- Paper
- Pencil
- 3 x 5 cards
- Ready, Set, Hop! Record Sheet, page 56

Activity 1

Form groups of two children and give each child 10 Unifix Cubes of 3 different colors. As suggested, reread the story to the children. One child will represent Moe and the other, Matty. For each segment, have children use their cubes to represent the number of jumps. They should change colors of cubes for each segment as illustrated below.

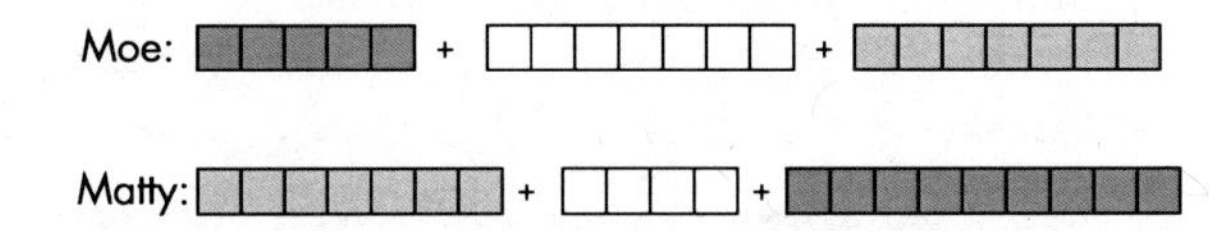

Activity 2

Give each child a plastic cup and an appropriate number of two-color counters (depending upon what facts are being taught). Prepare a three-column sheet similar to that illustrated below. Have children toss the counters on their desk, write the corresponding addends in the appropriate column, and write the addition number sentence. After a given time, discuss class findings. Discuss the commutative property for addition of whole numbers.

Red Counters	White Counters	Number Sentence
2	4	2 + 4 = 6

Activity 3

Fill paper bags with 10 Unifix Cubes of three different colors. Distribute the bags to pairs of children, along with a Ready, Set, Hop! Record Sheet illustrated below. Children take turns reaching into the bag without looking and grabbing a handful of cubes. They record the number of cubes of each color, and then write the corresponding number sentence with the three addends. Continue the activity until each child has 10 number sentences.

Red Cubes	Blue Cubes	Green Cubes	Number Sentence
2	3	5	2 + 3 + 5 = 10

Activity 4
Have a hopping contest in the classroom, hallway, or playground. Use masking tape or gray tape to designate a starting line and finish line. Give each child a class roster to record the number of hops for each classmate. Once all children have finished, have children analyze the data.

- Who had the fewest number of hops?
- Who had the most number of hops?

Depending on grade level, have children make frequency bar graphs for the data. Discuss the graphs. What is the mode (most frequent number of hops)? If appropriate, what is the mean? What is the median? What is the range?

Activity 5
Have children use Unifix Cubes in bars of 10 to measure the length (width) of the classroom. Each time a bar is added, have children write the corresponding number sentence.

Activity 6
Place plastic coins (pennies, nickels) in a paper bag. Have children grab a handful of coins and find the total amount of money.

Writing and Communicating
Have children write a "Ready, Set, Hop!" story about kangaroos, grasshoppers, or crickets, illustrating the number sentences used in the story.

Assessment
Present children with addition number sentences on cards such as the following:

3 + 5 + ☐ = 12   4 + ☐ + 2 = 9   ☐ + 1 + 5 = 10

Have children find the missing addends, using Unifix Cubes if necessary.

Notes:

RODEO TIME

Story Summary

Katie and Cameron help their Uncle Cactus Joe get ready for a rodeo show by doing odd jobs for him. They set schedules for themselves in order to be on time; however, various situations always cause them to be late. They find out how important it is to keep on a schedule.

New York: Harper Collins Publishers, 2006 ISBN: 0-06-055779-6

Concepts or Skills

- Scheduling

Objectives

- Calculate elapsed time
- Construct a schedule
- Construct a timeline

Materials Needed

- Paper
- Pencil
- 3 x 5 cards
- Rodeo Time Cards, pages 57-58

Activity 1

Give each child two or three 3 x 5 inch cards. Have them write their morning schedule on the cards (waking time until school arrival). Compare and discuss children's schedules.

Activity 2

Have children estimate how long it takes from them to ride a bus to school from home. Have them write down the exact time the bus picks them up and what time the bus arrives at school. Were their estmates close? How far off were they? Children can make a bar graph of how long it takes for each classmember to get to school.

Activity 3

Have children research how doctors and dentists schedule their patients.

Activity 4

Have students make their own TV schedule for one week. Have them list TV shows they watch.

Have them bring their schedules to class to compare with schedules of other children. Then, have children answer the following questions:

a. How many hours of TV do you watch per week?
b. Which classmember watches the most or least amount of TV weekly.
c. How many classmembers watch the same TV show each week?
d. How many children do not watch TV on a certain day?

Have children graph class information.

Activity 5

Have children do research about rodeos. Have them find information about the average time a cowboy rides a bull.

Activity 6

Make copies of the Rodeo Time Cards. Cut them apart and distribute a set to groups of two or three children. Have children order the times from earliest to latest time.

Activity 7

Make transparencies of the Rodeo Time Cards. Place two cards on the overhead projector and have children determine the elapsed time.

Make copies of the Rodeo Time Cards. Cut them apart and distribute a set to groups of two or three children. Have children place the cards face down in a pile. On a turn, a child draws two cards and finds the elapsed time.

Writing and Communicating

Have children write about how a schedule helps a person.

Assessment

Present children with a tangled list of schedule times for a particular event. Have children organize the list according to times.

Internet Links

www.americancowboy.com

www.pbrnow.com

Notes:

ROOM FOR RIPLEY

Story Summary

Carlos needs to determine how much water he will need for his fish, Ripley and Wiggles, in their fish bowl. He uses different units of measure for the water: cup, pint, half-gallon and gallon.

New York: Harper Collins Publishers, 1999 ISBN: 0-06-027620-7

Concepts or Skills

- Liquid measurement with cups, pints, quarts, half-gallons, gallons, liters

Objectives

- Estimate liquid measure with reasonable accuracy
- Measure a given liquid amount
- Describe the relationships among various liquid measures, both English and metric

Materials Needed

- Fish bowl
- Measuring cups of different sizes
- Pint, half-gallon and gallon containers
- Liter containers
- Paper

Activity 1

Show children an empty fish bowl. Have them estimate how much water would be needed to fill the bowl if they used various units. Discuss their estimates. Then fill the bowl using each of the measuring units.

Activity 2

Have children make a mobile that represents the following relationships: cups to pints to quarts to half-gallons to gallons.

Activity 3

As a homework assignment, have children complete a chart titled "What Comes in Different Containers?" Discuss the charts in class.

Activity 4

Have children work in cooperative groups to determine how much water, soda, juice, or milk is used for lunch by the class or by the school. How many cups? Pints? Quarts? Half-gallons? Gallons? Liters?

Writing and Communicating

Have children write on topics such as:

- Why do we need different containers for measuring?
- Where did the names come from?
- The Metric Way
- Barrels and Hogsheads

Assessment

Have children complete a chart showing the numerical relationships among all the liquid measurement units.

Have children estimate the number of cups or pints in a particular container.

Notes:

SAFARI PARK

Story Summary

Grandpa takes his five grandchildren to a new amusement park. Different rides at the park cost 1, 2, 4, or 6 tickets and snacks at the park cost 1 ticket. Each grandchild is given 20 tickets for the outing, but Paul forgets his tickets at home. Grandpa asks the other 4 children to take Paul on one ride. Before they spend any tickets, Grandpa asks them to decide what rides they want to go on so they won't run out of tickets. Chad, Alicia and Abby make initial choices that result in tickets being left over and they need to find the unknown amount so they know what additional ride(s) they can take. Patrick makes choices that exceed 20 tickets and has to decide what ride(s) to eliminate. The Terrible Tarantula is the one ride Paul really wants to take but it costs 6 tickets and none of the children decide to spend that many tickets for one ride. Paul ends up winning 36 tickets on a rock toss game and everyone enjoys the Terrible Tarantula, including Grandpa.

New York: Harper Collins Publishers, 2002 ISBN: 0-06-446245-5

Concepts or Skills

- Finding unknowns
- Patterning

Objectives

- Model combinations of addends that sum to 20
- Solve for an unknown quantity in a number sentence
- Complete a table and determine patterns for various combinations of addends

Materials Needed

- Unifix Cubes
- Poster paper
- Safari Park Sheet, page 59

Activity 1

While reading the story, have students use twenty Unifix Cubes to model how each child decides to spend his or her tickets. As the story continues, have students make new arrangements of the cubes as suggested by the story. Discuss the total number of rides the children are able to take with their twenty tickets, including the ride they take with Paul.

Activity 2

Divide the class into small groups. Distribute twenty Unifix Cubes to each student or group. Have each group explore different ways in which they could spend their twenty tickets. Have students write a number sentence for their ride selections as modeled in the story.

Activity 3

Distribute a Safari Park sheet for using twenty tickets. Have students investigate the combinations of rides, games, and treats possible for using all twenty tickets and record the results in the table. There are many combinations. Post several sheets on the bulletin board. Over a period of days, let students post different combinations.

Activity 4

Have five students role play the five grandchildren. Give twenty Unifix Cubes to each of four children. Pose the question of sharing the total number of

Unifix Cubes (80) among the five. How many would each student have? (16) The concept of averages could be further explored if additional students join and/or leave the group.

Writing and Communicating

Have students write their own stories about being given a fixed amount of something (money, candies, computer time) and deciding how to spend what they were given.

Assessment

Specify a given number of tickets to start and the number of tickets for a variety of rides. Have students write number sentences for different ways to spend their tickets.

Internet Links

www.gouchercenter.edu/jcampf/patterns.htm

www.earlyalgebra.terc.edu/

Notes:

SHARK SWIMATHON

Story Summary

Gill, Stripes, Tiny, Flip, Flap and Fin are sharks that want to go to the state swim camp. However, they have no money. In the local newspaper, the Ocean City Bank has offered to send any swim team to camp if they swim 75 laps by the end of the month.

All six sharks join in swimming laps. Their coach calculates the remaining number of laps at the end of each day. As the story continues, Gill gets hurt and is unable to swim any more laps. The remaining sharks must now swim extra laps to make up for his absence. In the end, the sharks complete their 75 laps and get to go to swim camp.

New York: Harper Collins Publishers, 2001 ISBN: 0-439-36572-4

Concepts or Skills

- Subtraction of two-digit numbers with and without regrouping
- Addition with multiple addends

Objectives

- Subtract multiplace numbers with and without regrouping
- Write another name for a number

Materials Needed

- Unifix Cubes
- Two-color counters
- Plastic cup
- Paper and pencils
- 3 x 5 cards
- Shark Attack Sheet, page 60
- Shark Base 10 Pieces, page 61

Activity 1

The sharks must swim 75 laps in four days. Have children work in cooperative groups and list as many ways as they can four addends that have a sum of 75. Children may want to use calculators or manipulatives to help with this activity.

Here are some examples:

$30 + 40 + 4 + 1 = 75$

$20 + 20 + 20 + 15 = 75$

$70 + 3 + 1 + 1 = 75$

Have children share some of their number sentences by writing them on the overhead or on the chalkboard.

Activity 2

Using the lap information from the story, have children determine which shark swam the most laps and which shark swam the least number of laps.

Depending upon grade level, have children determine the average number of laps each shark swam before Gill was hurt.

Activity 3

Have children work in cooperative groups. Give each group one six-sided number cube. Children write the number 75 (or some designated number) on their paper. Each child rolls the cube and subtracts the number showing from 75. Play continues until one player reaches zero. A player must exactly hit 0.

If a number cannot be subtracted, the player loses that turn.

As a variation, give each group two number cubes. On a turn, a child finds the sum of the two numbers and subtracts the result from 75. The game continues until a player reaches zero.

Activity 4

Distribute a Shark Attack Sheet to each child. The sheet has the numbers 1 through 10 written four times. Write a two-digit number on the overhead or chalkboard, and have the children write this number in the box on the Shark Attack Sheet. Tell children they must use three of the numbers to make a sum equal to the number that you have written. For example, if the number is 21, then 4 + 7 + 10 = 21. If the number is 15, then 4 + 5 + 6 = 15.

Note that there is more than one way to make several of the numbers. Numbers can only be used once.

Discuss the sums that children write. What is the greatest two-digit sum that can be found with three addends?
27 = 8 + 9 + 10
What is the least two-digit sum that can be written with three addends?
10 = 1 + 2 + 7

As a variation, have children use four addends. Also, increase the numbers used to 1 through 20.

Activity 5

Have children model the subtraction algorithm using Base 10 blocks. Write a problem on the overhead or chalkboard involving subtracting a two-digit number from a two-digit number. Begin with problems where no regrouping is necessary, then extend to problems with regrouping. Have children write the corresponding steps on paper. A page of Base 10 pieces are included.

Extend the activity to problems involving three digits.

Activity 6

Have children do research about the length of a standard swimming pool. Then have them determine the total number of meters that the sharks swam as they completed 75 laps

Writing and Communicating

Have children write about the different types of sharks in the story, giving various measurements (length, weight).

Assessment

Have children model a multiplace subtraction problem using Base 10 blocks.

Use Shark Attack as an assessment tool for addition with multiple addends.

Internet Links

www.animalnetwork.com

www.jackhanna.com

Notes:

SLUGGER'S CAR WASH

Story Summary

The 21st Street Sluggers need new tee shirts for their team. They decide to earn some money by having a car wash. Their parents would match the amount they made at the car wash. One member of the team, CJ, does not participate in washing cars, but merely collects money the entire day. His friends notice and appropriately soak him.

New York: Harper Collins Publishers, 2002 ISBN: 0-06-028920-1

Concepts or Skills

- Making change

Objectives

- Make change
- Determine the number of ways to make change for a given amount of money
- Compute decimal equivalents of rational numbers

Materials Needed

- Manila paper
- Plastic coins
- Paper and pencils
- Calculator
- Supermarket ads
- Coupons
- Fake discount cards for a store
- 3 x 5 cards
- Change Charts, page 62

Activity 1

The 21st Street Sluggers charged $3.50 for each car wash. Have children calculate how many cars they would need to wash in order to make the entire $100.00? Change the price to $4, $4.50 and $5. Have children calculate the results for the new prices.

Have children construct graphs for each price, plotting points such as (1, $4), (2, $8), . . . , (25, $100). Discuss the graphs for each. Note the steepness of the plots as the price increases.

Activity 2

CJ kept a record of how much money team members gave to purchase supplies to start the car wash. Have children calculate the following using plastic coins:

- How many quarters would it take to make $10.00 or $20.00?
- How many dimes would it take to make $5.00 or $10.00?
- How many nickels would it take to make $5.00?

Activity 3

Determining the number of ways to make change for various amounts of money is a powerful number theory problem that provides many extensions.

Distribute Change Charts to groups of two or three children. Indicate an amount of money for children to change using various coins. Post results on a bulletin board. Have children discuss their results.

If only one denomination of coin can be used to make change (1 dime = 2 nickels, 1 dollar = 4 quarters), then the number of combinations is limited; otherwise, results like the following are possible.

Amount = 10¢ Coins		
1¢	5¢	10¢
10	0	0
5	1	0
0	2	0
0	0	1

Amount = 25¢ Coins			
1¢	5¢	10¢	25¢
25	0	0	0
20	1	0	0
15	2	0	0
15	0	1	0
10	3	0	0
10	1	1	0
5	4	0	0
5	2	1	0
5	0	2	0
0	5	0	0
0	3	1	0
0	1	2	0
0	0	0	1

Activity 4

Give each child a supermarket ad and a calculator. Tell children they each have $20.00 to spend. Let them determine what they could buy with the $20.00. Have them list each item and how much it would cost? Have them determine how much change they would receive after they have gone shopping. Once they have finished, have different children count out the appropriate change.

For older students, do the same activity, but let them also determine the following:

- Discounts using coupons.
- Discounts using the store's discount card.
- Sales tax for those items that are taxed.

Activity 5

This activity is designed for children who have studied decimals. It is also appropriate to use as an introductory activity for calculators. Tell each child that they are playing for a major league baseball team and they each have been at bat 50 times. Create a set of cards that indicate a number of hits (5 to 20 with duplicates). Have children draw a card from the deck and determine their batting averages (hits divided by 50 at bats). Discuss the various batting averages.

Activity 6

This activity is designed as a weekend project. Prepare a list of common non-food items such as those for a car wash and have children go to a store in the area to find out how much the items cost. Discuss class findings. If possible, have children use the Internet to do comparison shopping for the items.

Activity 7

Set up a play bank with $20.00 in different bills and coins in the bank. On a 3 x 5 card, write different amounts of money that each child would give to the banker to pay for the car wash. The teacher would be the banker. Each child brings his/her card to the banker. The children will proceed to tell the banker how much money they would receive back in change. Children should tell the banker what coins or bills to use.

Writing and Communicating

Have children write about the "Bad Luck Two-dollar Bill."

Have children write about coins and bills from other countries such as Canada and Mexico.

Assessment

Give each child a card indicating an amount of money to spend and the amount spent. Have the child correctly make change using plastic coins and paper bills.

Have children determine the number of ways to make change for a particular amount of money and designated coins.

Internet Links

www.amazon.com

www.salescircular.com

www.statefarm.com/kidstuf/kidstuf.htm

Notes:

TOO MANY KANGAROO THINGS TO DO

Story Summary

On his birthday, Kangaroo is lonely because none of his animal friends will play with him. As he visits each of his platypus, emu, koala and dingo friends, he finds they all have too many things to do and cannot play with him. Little does Kangaroo realize that his friends are doing all the things for a surprise birthday party.

New York: Harper Collins Publishers, 1996 ISBN: 0-06-025883-7

Concepts or Skills

- Multiplication
- Multiples
- Patterning

Objectives

- Skip count by 2s, 3s and 4s
- Find basic multiplication products from 1 x 1 through 4 x 4
- Determine terms in a sequence of counting numbers
- Model basic multiplication facts using an array

Materials Needed

- Unifix Cubes
- 2 cm grid paper
- Dice
- Shade the Square! Grid, page 63
- Multiply and Add! Score Sheets, page 64

Activity 1

As suggested in the book, have students list things that come or occur in various counting numbers. Generally, the counting numbers involved are 1 through 6, 8, 10, 12 and 24.

Use the activity as a home connection assignment. Have students share their lists in class.

Activity 2

Each sequence of numbers presented in the book provides students with an opportunity to study patterns and determine additional terms in the sequence. Have students determine the first ten terms for each sequence.

- 1, 2, 3, 4, 5, 6, 7, 8, 9, 10
- 2, 4, 6, 8, 10, 12, 14, 16, 18, 20
- 3, 6, 9, 12, 15, 18, 21, 24, 27, 30
- 4, 8, 12, 16, 20, 24, 28, 32, 36, 40

Continue with other starting numbers 5 through 10.

Provide other number sequences such as the following:

- 3, 7, 11, 15, 19, 23, 27, 31, 35, 39
- 5, 8, 11, 14, 17, 20, 23, 26, 29, 32
- 32, 30, 28, 26, 24, 22, 20, 18, 16, 14, 12
- 0, -2, -4, -6, -8, -10, -12, -14, -16, -18

Activity 3

Play Shade the Square! Divide the class into groups of two. Distribute a Shade the Square! grid to each student. Give each group a pair of dice and 36 Unifix

Cubes. On a turn, a student tosses the pair of dice and writes a corresponding number sentence, a x b = c or b x a = c.

Have the students model the sentence using the cubes, and then shade the corresponding array on the Shade the Square! grid. Shaded areas cannot overlap. The first player who cannot shade an array loses the game.

For a quicker game, have students shade the same grid. Other rules remain the same.

Activity 4

Play Multiply and Add! Divide students into groups of three or four. Give each group 40 Unifix Cubes and a pair of dice. Give each student a Multiply and Add! score sheet. Taking turns, each student tosses the dice and uses Unifix Cubes to model the problem as an array. Once the product is found, the student records it on his/her score sheet and the next student tosses the dice. After five rounds, each student finds the sum of the five products. The student with the greatest total is the winner.

Writing and Communicating

Have each student create his/her own story involving animals and certain tasks to perform.

Have students respond to "Why Multiplication is Quicker Than Addition."

Assessment

Give students a sequence of numbers with missing terms. Have them determine the missing numbers.

Internet Links

www.aplusmath.com

www.edu4kids.com

Notes:

TREASURE MAP

Story Summary

The Elm Street Kids Club finds a buried treasure map. They follow directions on the map to find the treasure. It is a time capsule with different items buried by a kids club 50 years ago.

New York: Harper Collins Publishers, 2004 ISBN: 0-06-028036-0

Concepts or Skills

- Map reading
- Scales
- Proportions
- Measurement
- Directions
- Coordinates

Objectives

- Follow a given set of map directions
- Read a ruler to the nearest millimeter
- Read a ruler to the nearest 1/8 inch
- Solve proportions
- Graph data
- Estimate quantities

Materials Needed

- Small U.S. maps with major cities marked and a scale given
- Mileage chart from an atlas

Activity 1

Have children make a time capsule and place current items from our time in the capsule. Bury the time capsule on school grounds. Have children in pairs make a map and give directions to find the capsule from various points. Discuss their map and directions. Will the starting point be there in 50 years? Will it be covered by some building?

Activity 2

Distribute a ruler and U.S. map to each child. Designate two cities marked on the map. Have children measure the straight line distance between the two cities. Discuss their measurements before proceeding. Then have children use the given map scale to find the distance between the cities.

Use an atlas mileage chart to check the distance.

Have children go to an online site such as Mapquest to find directions and mileage.

Activity 3

Have children draw maps to get from their classroom to the cafeteria or gym. Use a child's steps for measurement units. If a trundle wheel for measurement is available, have a group of children use it to determine more accurate measurements for distances.

Activity 4

Have children draw maps to get from their homes to school. If possible, have parents give mileage units.

Activity 5

On a large wall map, have children mark their birthplaces with pins. Have them determine the distance in miles to the school/city location.

On a large wall map, have children mark the birthplaces of U.S. presidents. Have them determine the distance in miles to the school/city location.

Activity 6

Form small groups of two or three children. Have each group determine a starting point and ending point (treasure) in the classroom. Then have them write clues to get to the treasure.

Once directions have been written, have groups exchange their directions with other groups. Let children find the "treasure" using the directions.

Activity 7

Copy a local city map found in most telephone directories. Locate the school and designate a destination for children. Have them write directions to get from the school to the destination. Discuss directions. Did every child write the same direction? Who has the best directions? Why?

Writing and Communicating

Give each child a simple square grid "map" with certain designated sites. Have the child write or orally describe how to get from one point to the other.

Have children write about the following prompts. They must also provide mileage information in their writing.

- The Time Capsule I Found
- My Favorite Vacation
- If I Could Travel Anywhere in the U.S. (in the World)

Assessment

Give each child a small U.S. map with certain designated cities. Have the child measure the distance with a ruler and use the scale factor to determine the distance in miles or kilometers.

Internet Links

www.nationgeographic.com/maps

www.freetrip.com

www.mapsonus.com

Notes:

2 CM GRID PAPER

BLANK CIRCLE GRAPH

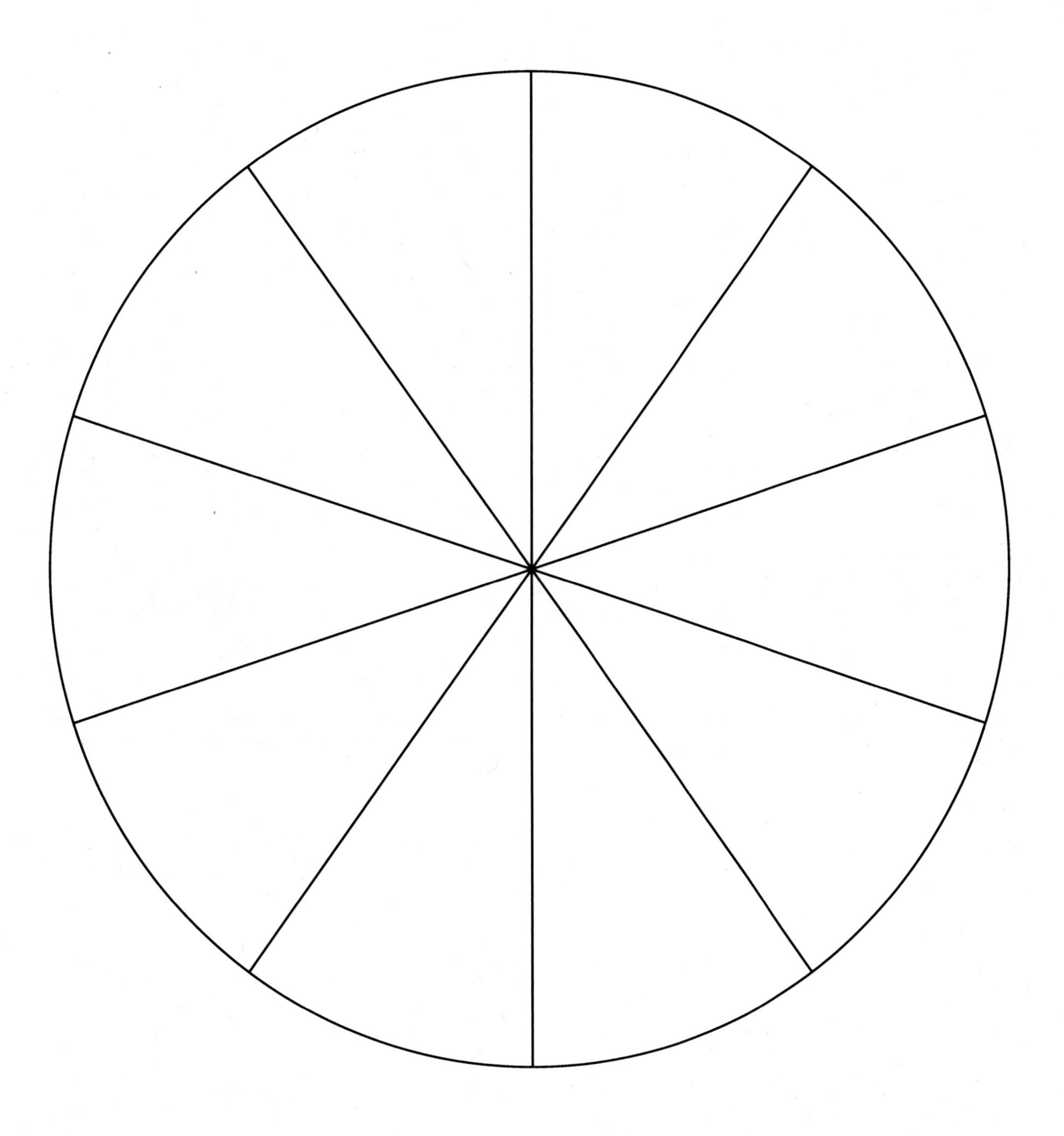

FRACTION CARDS

1/2	1/4	3/4
1/10	3/10	1/5
2/5	3/5	4/5
2/4	40/100	30/100
50/100	5/10	4/10

NAME GRAPH BLANKS

CLAM NUMBER LINES

-4	-3	-2	-1	0	1	2	3

	4	5	6	7	8	9	10

-4	-3	-2	-1	0	1	2	3

	4	5	6	7	8	9	10

-4	-3	-2	-1	0	1	2	3

	4	5	6	7	8	9	10

LESS THAN ZERO STORY GRAPH

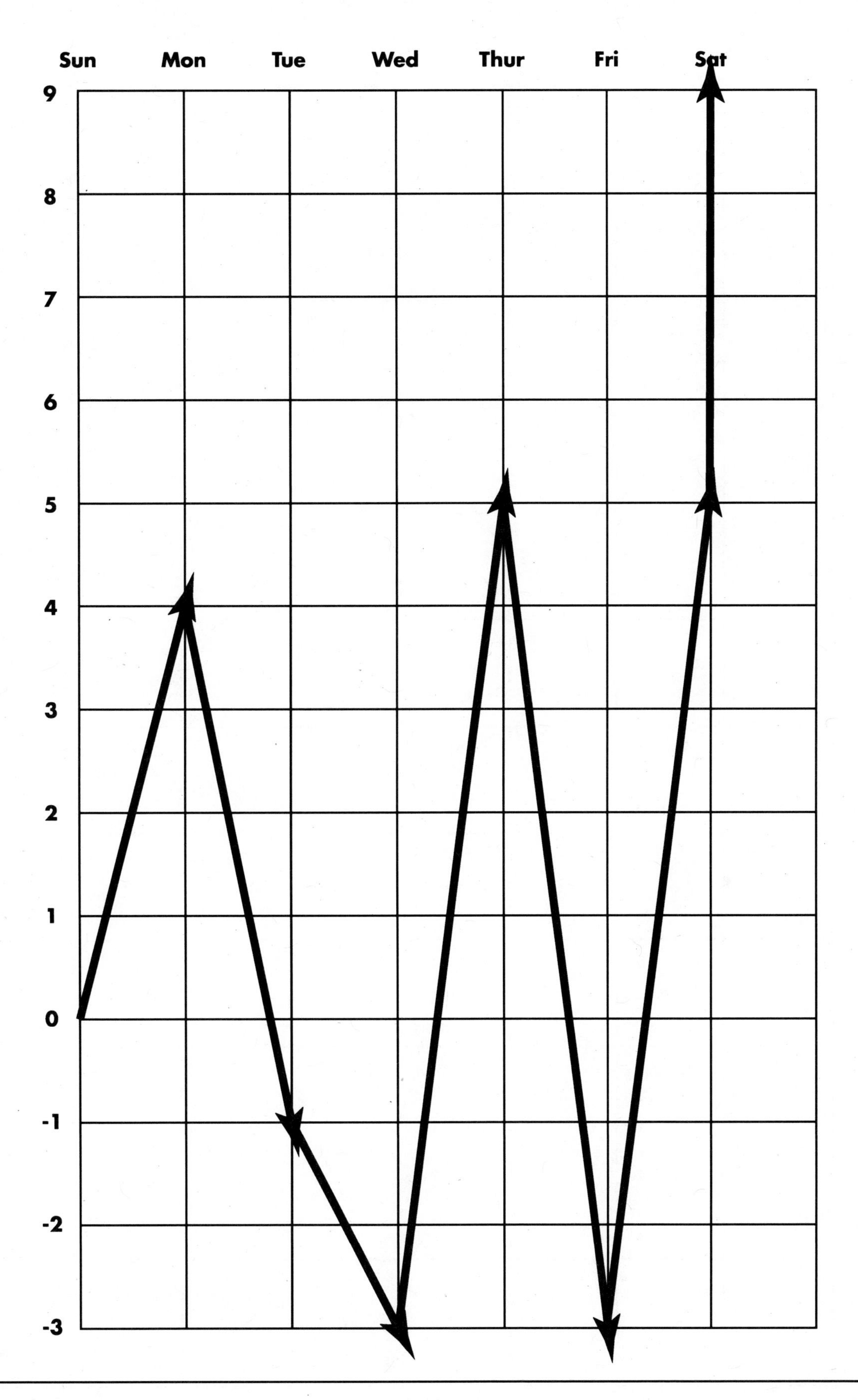

LESS THAN ZERO WEEK GRAPH

Sun	Mon	Tue	Wed	Thur	Fri	Sat

COIN MAT

Penny	Nickel	Dime	Quarter

READY, SET, HOP! RECORD SHEET

Cubes	Cubes	Cubes	Number Sentence

RODEO TIME CARDS, 1

8:00 A.M.	**9:00 A.M.**	**10:00 A.M.**
11:00 A.M.	**8:30 A.M.**	**9:30 A.M.**
10:30 A.M.	**11:30 A.M.**	**8:45 A.M.**
9:45 A.M.	**10:45 A.M.**	**11:45 A.M.**
12:00 P.M.	**1:00 P.M.**	**2:00 P.M.**
3:00 P.M.	**4:00 P.M.**	**5:00 P.M.**

RODEO TIME CARDS, 2

6:00 P.M.	12:30 P.M.	1:30 P.M.
2:30 P.M.	3:30 P.M.	4:30 P.M.
5:30P.M.	1:45 P.M.	2:45 P.M.
3:45 P.M.	4:45 P.M.	5:45 P.M.
6:30 P.M.	6:45 P.M.	7:00 P.M.
7:30 P.M.	7:45 P.M.	8:00 P.M.

SAFARI PARK SHEET

Name	Rides, Games, Treats					Total Tickets
	Jungle Kings 4 Tickets	Rhino Rides 2 Tickets	Monkey Games 1 Tickets	Tiger Treats 1 Tickets	Terrible Tarantula 4 Tickets	

SHARK ATTACK SHEET

1 2 3 4 5 6 7 8 9 10

1 2 3 4 5 6 7 8 9 10

1 2 3 4 5 6 7 8 9 10

1 2 3 4 5 6 7 8 9 10

SHARK BASE 10 PIECES

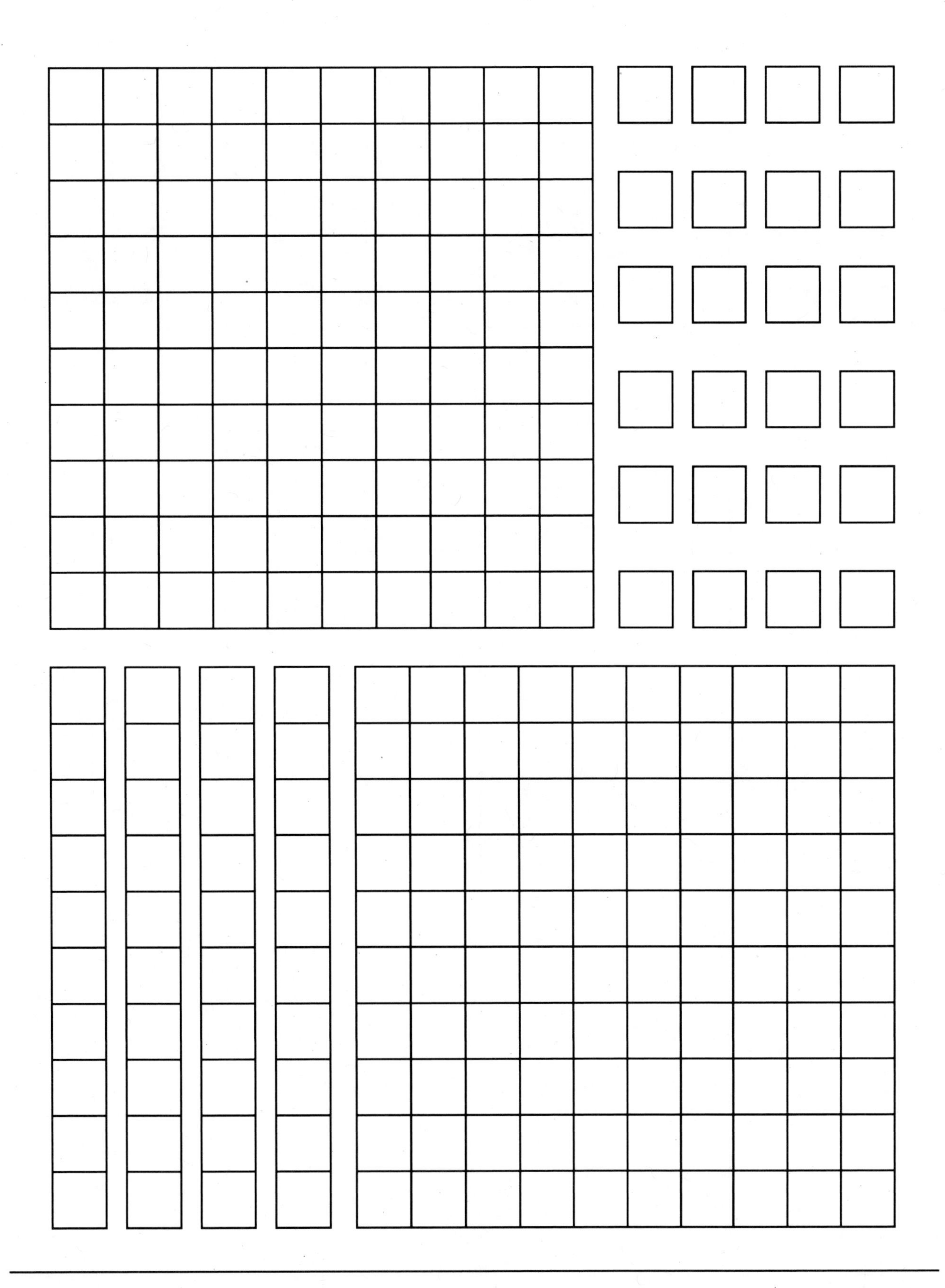

CHANGE CHART

Penny	Nickel	Dime	Quarter

SHADE THE SQUARE GRID

Toss a pair of dice. Write a multiplication sentence for your toss.
Shade the array below or place Unifix Cubes on the grid to model your array.

MULTIPLY AND ADD!

MULTIPLY AND ADD!	
Round	Number Sentence a x b = c
1	
2	
3	
4	
5	
Total	

MULTIPLY AND ADD!	
Round	Number Sentence a x b = c
1	
2	
3	
4	
5	
Total	

MULTIPLY AND ADD!	
Round	Number Sentence a x b = c
1	
2	
3	
4	
5	
Total	

MULTIPLY AND ADD!	
Round	Number Sentence a x b = c
1	
2	
3	
4	
5	
Total	

MULTIPLY AND ADD!	
Round	Number Sentence a x b = c
1	
2	
3	
4	
5	
Total	

MULTIPLY AND ADD!	
Round	Number Sentence a x b = c
1	
2	
3	
4	
5	
Total	